100

things you should know about

PLANET EARTH

100

things you should know about

PLANET EARTH

Peter Riley

Consultant: Clive Carpenter

Mason Crest
Publishers

Mason Crest Publishers Inc.
370 Reed Road, Broomall, PA 19008
(866) MCP-BOOK (toll free)
www.masoncrest.com
This edition first published in 2003

First published in 2001, Miles Kelly Publishing,
Bardfield Centre, Great Bardfield, Essex, CM7 4SL, U.K.
Copyright © Miles Kelly Publishing 2001, 2003

2 4 6 8 10 9 7 5 3 1

Library of Congress Cataloging-in-Publication Data on file
at the Library of Congress

ISBN 1-59084-454-8

Editorial Director: Anne Marshall
Editors: Amanda Learmonth, Jenni Rainford
Design: Angela Ashton, Joe Jones
Indexing, Proof Reading: Lynn Bresler
Americanization: Sean Connolly

Printed in China

ACKNOWLEDGMENTS
The publishers would like to thank the following artists who have
contributed to this book:

Jim Channell
Kuo Kang Chen
Richard Draper
Chris Forsey
Mike Foster/Maltings Partnership
Studio Galante
Alan Hancocks

Kevin Maddison
Janos Marffy
Terry Riley
Martin Sanders
Mike Saunders
Rudi Vizi

Cartoons by Mark Davis at Mackerel

www.mileskelly.net
info@mileskelly.net

Contents

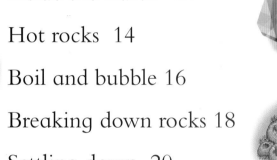
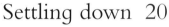
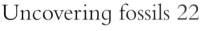
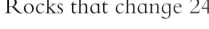

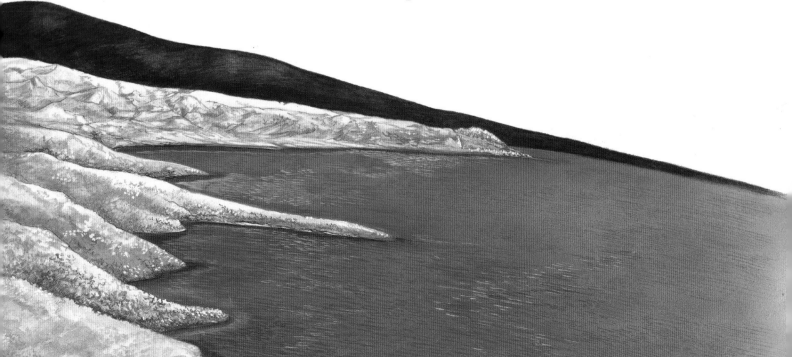

The speedy space ball

1 The Earth is a huge ball of rock moving through space at nearly 2mi/sec (3,000m/sec). It weighs six billion, million, million tons. Up to two thirds of the Earth's rocky surface is covered by water—this makes the seas and oceans. Rock that is not covered by water makes the land. Surrounding the Earth is a layer of gases called the atmosphere (air). This reaches about 400mi (700km) from the Earth's surface—then space begins.

Where did Earth come from?

2 **The Earth came from a cloud in space.**
Scientists think the Earth formed from a huge cloud of gas and dust around 4.5 billion years ago. A star near the cloud exploded, making the cloud spin. As the cloud spun around, gases gathered at its center and formed the Sun. Dust whizzed around the Sun and stuck together to form lumps of rock. In time the rocks crashed into each other to make the planets. The Earth is one of these planets.

5. The Earth was made up of one large piece of land, now split into seven chunks known as continents

▶ Clouds of gas and dust are made by the remains of old stars that have exploded or simply stopped shining. It is here that new stars and their planets form.

1. Cloud starts to spin

4. Volcanoes erupt, releasing gases, helping to form the first atmosphere

3. The Earth begins to cool and a hard shell forms

3 **At first the Earth was very hot.** As the rocks crashed together they warmed each other up. Later, as the Earth formed, the rocks inside it melted. The new Earth was a ball of liquid rock with a thin, solid shell.

2. Dust gathers into lumps of rock, which form a small planet

4 Huge numbers of large rocks called meteorites crashed into the Earth. They made round hollows on the surface. These hollows are called craters. The Moon was hit with rocks at the same time. Look at the Moon with binoculars—you can see the craters that were made long ago.

▼ Erupting volcanoes and fierce storms helped form the atmosphere and oceans. These provided energy that was needed for life on Earth to begin.

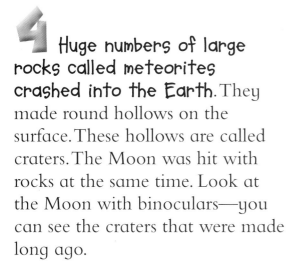

► The Moon was also hit by rocks in space, and these made huge craters, and mountain ranges up to 16,000ft (5,000m) high.

5 The oceans and seas formed as the Earth cooled down. Volcanoes erupted, letting out steam, gases, and rocks from inside the Earth. As the Earth cooled, the steam changed to water droplets and made clouds. As the Earth cooled further, rain fell from the clouds. It rained for millions of years to make the seas and oceans.

I DON'T BELIEVE IT!

Millions of rocks crash into Earth as it speeds through space. Some larger ones may reach the ground as meteorites.

In a spin

6 **The Earth is like a huge spinning top.** It continues to spin because it was formed from a spinning cloud of gas and dust. It does not spin straight up like a top but leans a little to one side. The Earth takes 24 hours to spin around once. We call this period of time a day.

7 **The Earth's spinning makes day and night.** Each part of the Earth spins toward the Sun, and then away from it every day. When a part of the Earth is facing the Sun it is daytime there. When that part is facing away from the Sun it is nighttime. Is the Earth facing the Sun or facing away from it where you are?

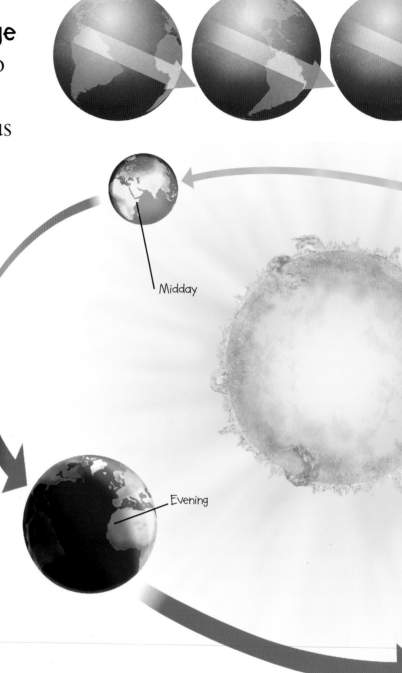

Midday

Evening

◀ If you were in space and looked at the Earth from the side, it would appear to move from left to right. If you looked down on Earth from the North Pole, it would seem to be moving anticlockwise.

8 **The Earth spins around its Poles.** The Earth spins around two points on its surface. They are at opposite ends of the Earth. One is on top of the Earth. It is called the North Pole. The other is at the bottom of the Earth. It is called the South Pole. The North and South Poles are so cold, that they are covered by ice and snow.

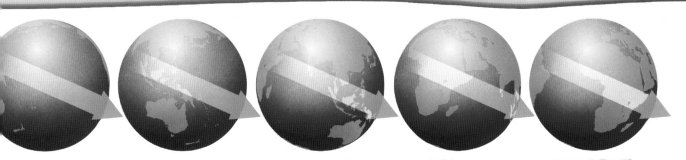

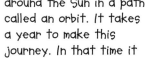

▲ The Earth moves around the Sun in a path called an orbit. It takes a year to make this journey. In that time it spins round 365 and a quarter times.

Morning

Night

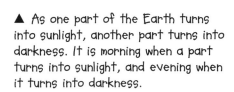

▲ As one part of the Earth turns into sunlight, another part turns into darkness. It is morning when a part turns into sunlight, and evening when it turns into darkness.

MAKE A COMPASS

A compass is used to find the direction of the North and South Poles.

You will need:

bowl of water   piece of wood
bar magnet   real compass

Place the wood in the water with the magnet on top. Make sure they do not touch the sides. When the wood is still, check the direction the magnet is pointing in with your compass, by placing it on a flat surface. It will tell you the direction of the North and South Poles.

9 **The spinning Earth acts like a magnet.** At the center of the Earth is liquid iron. As the Earth spins, it makes the iron behave like a magnet with a North and South Pole. These act on the magnet in a compass to make the needle point to the North and South Poles.

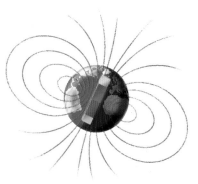

▲ These lines show the pulling power of the magnet inside the Earth.

Inside the Earth

10 **There are different parts to the Earth.** There is a thin, rocky crust, a solid middle called the mantle, and a center called the core. The outer part of the core is liquid but the inner core is made of solid metal.

11 **At the center of the Earth is a huge metal ball called the inner core.** It is 1,500mi (2,500km) wide and is made mainly from iron, with some nickel. The ball has an incredible temperature of 10,800°F (6,000°C)—hot enough to make the metals melt. They stay solid because other parts of the Earth push down heavily on them.

12 **Around the center of the Earth flows a hot, liquid layer of iron and nickel.** This layer is the outer core and is about 1,300mi (2,200km) thick. As the Earth spins, the metal ball and liquid layer move at different speeds.

▼ If the Earth could be cut open, this is what you would see inside. It has layers inside it like an onion.

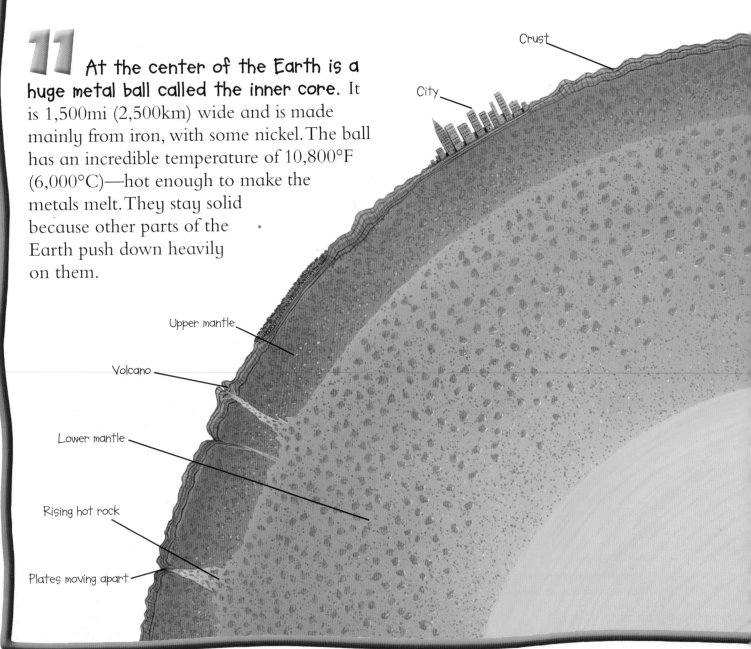

Crust

City

Upper mantle

Volcano

Lower mantle

Rising hot rock

Plates moving apart

13 The largest part of Earth is a layer called the mantle, which is 1,750mi (2,900km) thick. It lies between the core and the crust. Near the crust, the mantle is made of slow-moving rock. When you squeeze an open tube of toothpaste, the toothpaste moves a little like the rocks in the upper mantle.

14 The Earth's surface is covered by crust. Land is made of continental crust between 12 and 40mi (20 and 70km) thick. Most of this is made from a rock called granite. The ocean bed is made of oceanic crust about 5mi (8km) thick. It is made mainly from a rock called basalt.

15 The crust is divided into huge slabs of rock called plates. Most plates have land and seas on top of them but some, like the Pacific Plate, are mostly covered by water. The large areas of land on the plates are called continents. There are seven continents—Africa, Asia, Europe, North America, South America, Oceania, and Antarctica.

Outer core

Inner core

16 Very, very slowly, the continents are moving. Slow-flowing mantle under the crust moves the plates across the Earth's surface. As the plates move, so do the continents. In some places, the plates push into each other. In others, they move apart. North America is moving about 1in (2.5cm) away from Europe every year!

◄ There are gaps in the Earth's crust where hot rocks from inside can reach the surface.

Hot rocks

17 There are places on Earth where hot, liquid rocks shoot up through its surface. These are volcanoes. Beneath a volcano is a huge space filled with molten (liquid) rock. This is the magma chamber. Inside the chamber, pressure builds like the pressure in a can of soda if you shake it. Ash, steam, and molten rock called lava escape from the top of the volcano—this is an eruption.

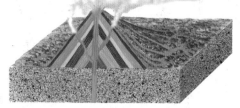

▲ These volcanoes are a shield volcano (top), a crater volcano (middle) and a cone–shaped volcano (bottom).

18 Volcanoes erupt in different ways and have different shapes. Most have a central tube called a pipe, reaching up to the vent opening. Some volcanoes have runny lava, like those in Hawaii. It flows from the vent and makes a domed shape called a shield volcano. Other volcanoes have thick lava. When they erupt, gases in the lava make it explode into pieces of ash. The ash settles on the lava to make a cone-shaped volcano. A caldera, or crater volcano, is made when the top of a cone-shaped volcano explodes and sinks into the magma chamber.

Cloud of ash, steam, and smoke

Layers of rocks from previous eruptions

Lava flowing away from vent

Huge chamber of magma (molten rock) beneath the volcano

Molten rock spreading out under the volcano and cooling down

20 **Hot rocks don't always reach the surface.** Huge lumps of rock rise into the crust and can become stuck. These are batholiths. The rock cools slowly and large crystals form. When the crystals cool, they form a rock called granite. In time, the surface of the crust may wear away and the top of the batholith appears above ground.

◄ When a volcano erupts, the hot rock from inside the Earth escapes as ash, smoke, flying lumps called volcanic bombs, and rivers of lava.

MAKE YOUR OWN VOLCANO
You will need:
bicarbonate of soda   plastic bottle
food coloring   vinegar   sand
Put a tablespoon of bicarbonate of soda in the plastic bottle. Stand the bottle in a tray and make a cone of sand around it. Put a few drops of red food coloring in half a cup of vinegar. Tip the vinegar into a pitcher then pour it into the bottle. In a few moments the volcano should erupt with red, frothy lava.

19 **There are volcanoes under the sea.** Where plates in the crust move apart, lava flows out from rift volcanoes to fill the gap. The hot lava is cooled quickly by the sea and forms pillow-shaped lumps called pillow lava.

Boil and bubble

21 A geyser can be found on top of some old volcanoes. If these volcanoes collapse, their rocks settle above hot rocks in the old magma chamber. The gaps between the broken rocks make a group of pipes and chambers. Rainwater seeps in, collecting in the chambers, where it is heated until it boils. Steam builds up, pushing the water through the pipes and out of a cone-shaped opening called a nozzle. Steam and water shoot through the nozzle, making a fountain up to 200ft (60m) high.

▲ Geysers are common in the volcanic regions of New Zealand in Oceania. In some areas they are even used to help make electricity.

22 In the ocean are hot springs called black smokers. They form near rift volcanoes, where water is heated by the volcanoes' magma chambers. The hot water dissolves chemicals in the rocks, which turn black when they are cooled by the surrounding ocean water. They rise like clouds of smoke from chimneys.

23 In a hot spring, the water bubbles gently to the surface. As the water is heated in the chamber, it rises up a pipe and into a pool. The pool may be brightly colored due to tiny plants and animals called algae and bacteria. These live in large numbers in the hot water.

◄ The chimneys of a black smoker are made by chemicals in the hot water. These stick together to form a rocky pipe.

24

Wallowing in a mud pot can make your skin soft and smooth. A mud pot is made when fumes break down rocks into tiny pieces. These mix with water to make mud. Hot fumes push through the mud, making it bubble. Some mud pots are cool enough to wallow in.

▼ Mud pot

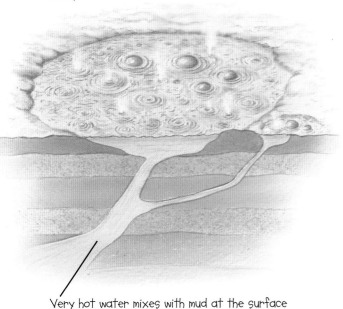

Very hot water mixes with mud at the surface

▲ The bubbles in a mud pot grow as they fill with fumes. Eventually they pop and the fumes escape into the air.

26

In Iceland, underground steam is used to make lights work. The steam is sent to power stations and is used to work generators to make electricity. The electricity then flows to homes and powers electrical equipment such as lights, televisions, and computers.

25

Steam and smelly fumes can escape from holes in the ground. These holes are called fumaroles. Since Roman times, people have used the steam from fumaroles for steam baths. The steam may keep joints and lungs healthy.

▼ Fumarole   Released steam

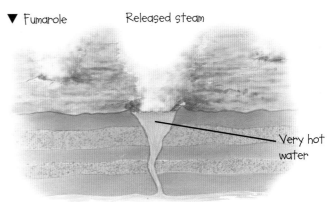

Very hot water

▲ Under a fumarole the water gets so hot that it turns to steam, then shoots upwards into the air.

MAKE A GEYSER
You will need:
pail   plastic funnel
plastic tubing

Fill a pail with water. Turn the plastic funnel upside down and sink most of it in the water. Take a piece of plastic tube and put one end under the funnel. Blow down the other end of the tube. A spray of water and air will shoot out of the funnel. Be prepared for a wet face!

Breaking down rocks

27 Ice has the power to make rocks crumble. In cold weather, rainwater gets into cracks in rocks and freezes. Water swells as it turns to ice. The ice pushes with such power on the rock that it opens up the cracks. Over a long time, a rock can be broken down into thousands of tiny pieces.

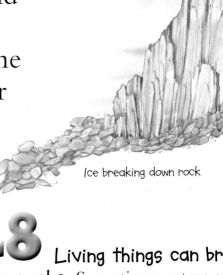

Ice breaking down rock

A tree root pushing its way through rock

28 Living things can break down rocks. Sometimes a tree seed lands in a crack in a rock. In time, a tree grows and its large roots smash open the rock. Tiny living things called lichens dissolve the surface of rocks to reach minerals they need to live. When animals, such as rabbits, make a burrow they may break up some of the rock in the ground.

29 Warming up and cooling down can break rocks into flakes. When a rock warms up it swells a little. When it cools, the rock shrinks back to its original size. After swelling and shrinking many times some rocks break up into flakes. Sometimes layers of large flakes form on a rock and make it look like onion skin.

▶ The flakes of rock break off unevenly and makes patterns of ridges on the rock surface.

30 Glaciers break up rocks and carry them away. Glaciers are huge areas of ice that form near mountaintops. They slide slowly down the mountainside and melt. As a glacier moves, some rocks are snapped off and carried along. Others are ground up and carried along as grit and sand.

▶ Snow falls on mountaintops and squashes down to make ice. The ice forms the glacier, which slowly moves down the mountainside until it melts.

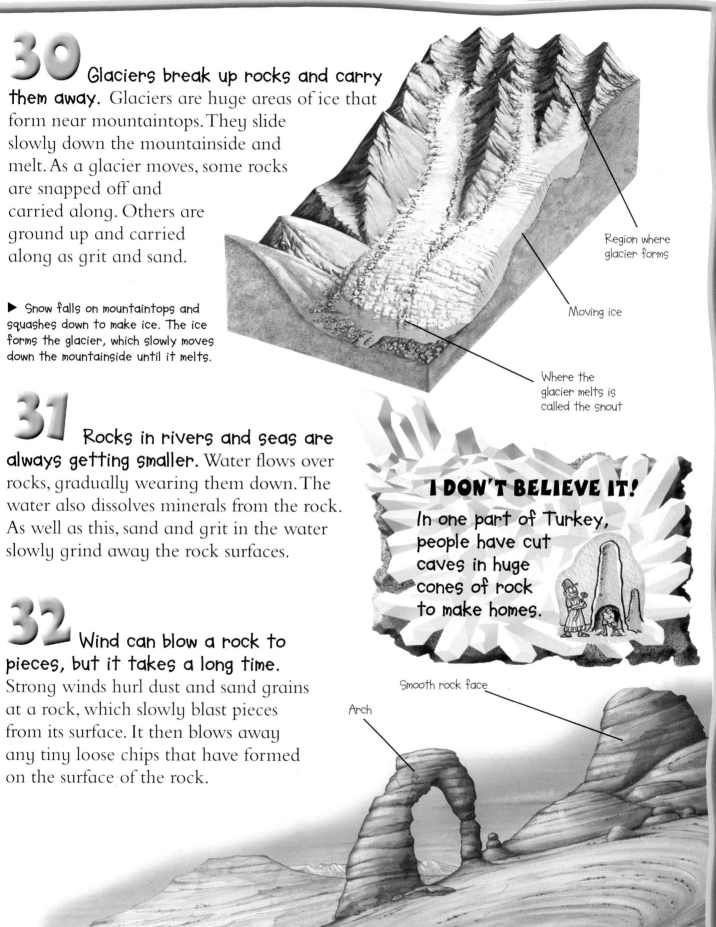

Region where glacier forms

Moving ice

Where the glacier melts is called the snout

31 Rocks in rivers and seas are always getting smaller. Water flows over rocks, gradually wearing them down. The water also dissolves minerals from the rock. As well as this, sand and grit in the water slowly grind away the rock surfaces.

I DON'T BELIEVE IT!

In one part of Turkey, people have cut caves in huge cones of rock to make homes.

32 Wind can blow a rock to pieces, but it takes a long time. Strong winds hurl dust and sand grains at a rock, which slowly blast pieces from its surface. It then blows away any tiny loose chips that have formed on the surface of the rock.

Smooth rock face

Arch

Settling down

33 Stones of different sizes can stick together to make rock. Thousands of years ago, boulders, pebbles, and gravel settled on the shores of seas and lakes. These have become stuck together to make a rock called conglomerate. At the foot of cliffs, broken, rocky pieces collected and stuck together to make a rock called breccia. The lumps in breccia have sharp edges.

▲ Pieces of rock can become stuck together by a natural cement to make a lump of larger rock, such as breccia.

34 Sandstone can be made in the sea or in the desert. When a thick layer of sand builds up, the grains are pressed together and cement forms. This sticks the grains together to make sandstone. Sea sandstone may be yellow with sharp-edged grains. Desert sandstone may be red with round, smooth grains.

▲ Natural cement binds grains of sand together to make sandstone.

35 If mud is squashed hard enough, it turns to stone. Mud is made from tiny particles of clay and slightly larger particles called silt. When huge layers of mud formed in ancient rivers, lakes, and seas, they were squashed by their own weight to make mudstone.

▶ Mudstone has a very smooth surface. It may be gray, black, brown, or yellow.

36
Limestone is made from seashells. Many kinds of sea animal have a hard shell. When the animal dies, the shell remains on the sea floor. In time, large numbers of shells build up and press together to form limestone. Huge numbers of shells become fossils.

▶ Limestone is usually white, cream, gray, or yellow. Caves often form in areas of limestone.

37
Chalk is made from millions of shells and the remains of tiny sea creatures. A drop of sea water contains many microscopic organisms (living things). Some of these organisms have shells full of holes. When these organisms die, the shells sink to the seabed and in time form chalk.

SEE BITS OF ROCK SETTLE
You will need:

sand   clay   gravel
plastic bottle

Put a tablespoon of sand, clay, and gravel into a bowl. Mix up the gravel with two cups of water then pour into a plastic bottle. You should see the bits of gravel settle in layers, with the smallest pieces at the bottom and the largest at the top.

I DON'T BELIEVE IT!
Flint is found in chalk and limestone. Thousands of years ago people used flint to make axes, knives, and arrowheads.

▲ Most chalk formed at the time of the dinosaurs, but chalk is forming in some places on the Earth today.

Uncovering fossils

38 **The best fossils formed from animals and plants that were buried quickly.** When a plant or animal dies, it is usually eaten by other living things so that nothing remains. If the plant or animal was buried quickly after death, or even buried alive, its body may be preserved.

▶ This is a fossil skull of *Tyrannosaurus rex*, a dinosaur that roamed the Earth around 70–65 million years ago.

39 **A fossil is made from minerals.** A dead plant or animal can be dissolved by water. An empty space in the shape of the plant or animal is left in the mud and fills with minerals from the surrounding rock. Sometimes, the minerals simply settle in the body, making it harder and heavier.

1. The trilobite lives on the ocean floor

2. The trilobite dies

3. The trilobite is covered by mud

4. The mud turns to stone

5. The fossil forms inside the stone

▲ Many fossils of trilobites, small ocean-dwelling creatures, have been found.

40 **Some fossils look like coiled snakes but are really shellfish.** These are ammonites. An ammonite's body was covered by a spiral shell. The body rotted away leaving the shell to become the fossil. Ammonites lived in the seas at the same time as the dinosaurs lived on land.

▲ When this ammonite was alive, tentacles would have stuck out of the uncoiled end of the shell.

41 **Dinosaurs did not just leave fossil bones.** Some left whole skeletons behind while others are known from only a few bones. Fossilized teeth, skin, eggs, and droppings have been found. When dinosaurs walked across mud they left tracks behind that became fossils. By looking at these, scientists have discovered how dinosaurs walked and how fast they could run.

42 **Electricity in your home may have been made by burning fossils.** About 300 million years ago the land was covered by forests and swamps. When plants died they fell into the swamps and did not rot away. Over time, their remains were squashed and heated so much that they turned to coal. Today, coal is used to work generators that make electricity.

I DON'T BELIEVE IT!
Some fossils of bacteria are three and a half billion years old.

▶ Coal was formed by trees and plants growing near water. When the trees died the waterlogged ground stopped them rotting away, and peat formed.

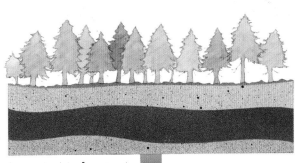

Dead trees are buried and squashed to form peat

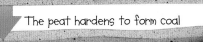

The peat hardens to form coal

Rocks that change

43 **When a rock forms in the crust it may soon be changed again.** There are two main ways this can happen. In one way, the rock is heated by hot rocks moving up through the crust. In another way the crust is squashed and heated as mountains form. Both of these ways make crystals in the rock change to form new types of rocks.

▶ Under the ground are layers of rock and some of them can be changed by heat.

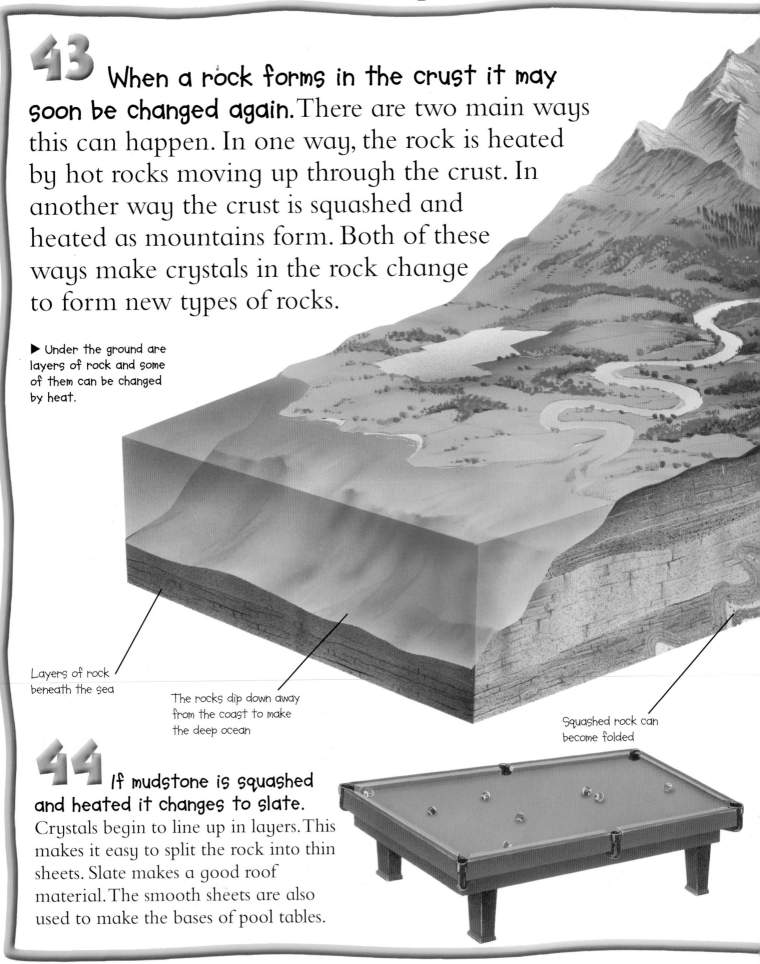

Layers of rock beneath the sea

The rocks dip down away from the coast to make the deep ocean

Squashed rock can become folded

44 **If mudstone is squashed and heated it changes to slate.** Crystals begin to line up in layers. This makes it easy to split the rock into thin sheets. Slate makes a good roof material. The smooth sheets are also used to make the bases of pool tables.

46 **Rock can become stripy when it is heated and folded.** It becomes so hot, it almost melts. Minerals that make up the rock form layers that appear as colored stripes. These stripes may be wavy, showing the way the rock has been folded. This rock is called gneiss (sounds like "nice").

Some hot rock travels to the surface through the pipe in a volcano

▲ The stripes in gneiss are formed by layers of different minerals.

Layers of rock away from the heat remain unchanged

Hot rock trapped in the crust can change the rock around it

45 **If limestone is cooked in the crust it turns to marble.** The shells that make up limestone break up when they are heated strongly and form marble, a rock that has a sugary appearance. The surface of marble can be polished to make it look attractive and it is used to make statues and ornaments.

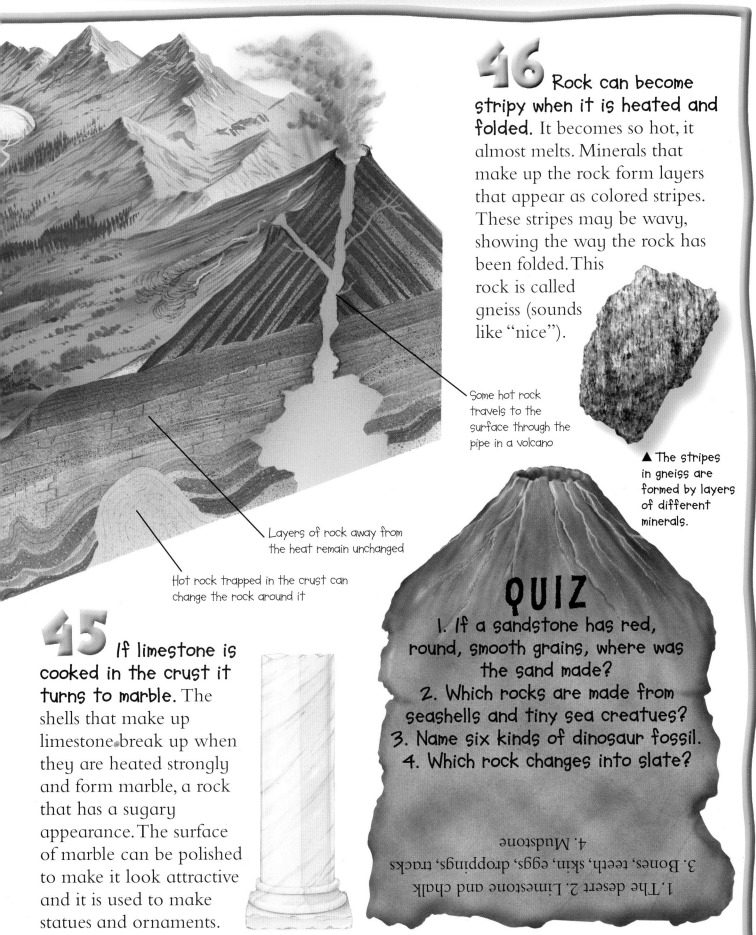

QUIZ
1. If a sandstone has red, round, smooth grains, where was the sand made?
2. Which rocks are made from seashells and tiny sea creatures?
3. Name six kinds of dinosaur fossil.
4. Which rock changes into slate?

1. The desert 2. Limestone and chalk 3. Bones, teeth, skin, eggs, droppings, tracks 4. Mudstone

Massive mountains

47 The youngest mountains on Earth are the highest. Highest of all is Mount Everest, which formed 15 million years ago. Young mountains have jagged peaks because softer rocks on the mountaintop are broken down by the weather. These pointy peaks are made from harder rocks that take longer to break down. In time, even these hard rocks are worn away. This makes an older mountain shorter and gives its top a rounded shape.

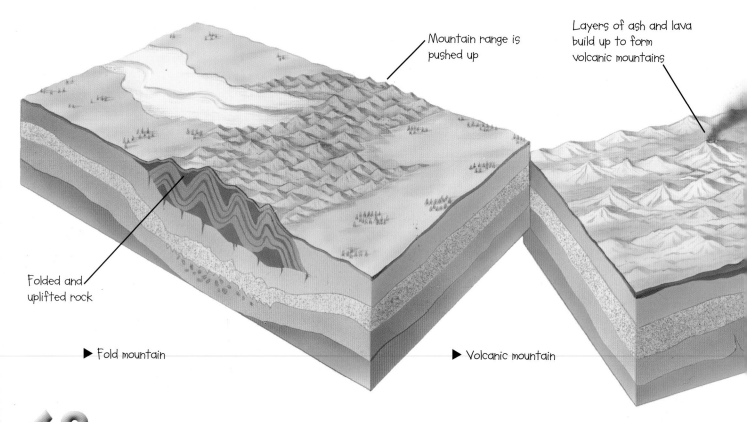

Mountain range is pushed up

Layers of ash and lava build up to form volcanic mountains

Folded and uplifted rock

▶ Fold mountain

▶ Volcanic mountain

48 When plates in the Earth's crust crash together, mountains are formed. When two continental plates crash together, the crust at the edge of the plates crumples and folds, pushing up ranges of mountains. The Himalayan Mountains in Asia formed in this way.

49 Some of the Earth's highest mountains are volcanoes. These are formed when molten rock (lava) erupts through the Earth's crust. As the lava cools, it forms a rocky layer. With each new eruption, another layer is added.

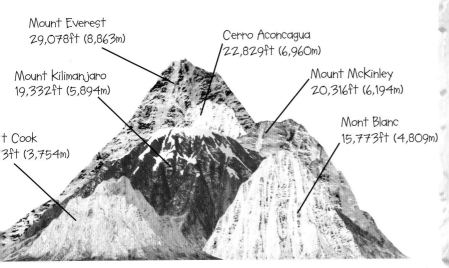

Mount Everest
29,078ft (8,863m)

Cerro Aconcagua
22,829ft (6,960m)

Mount Kilimanjaro
19,332ft (5,894m)

Mount McKinley
20,316ft (6,194m)

t Cook
3ft (3,754m)

Mont Blanc
15,773ft (4,809m)

▲ Mountains are the tallest things on Earth. Mount Cook is the smallest mountain shown here, and is still six times taller than the world's tallest man-made structure!

MAKE FOLD MOUNTAINS

Put a towel on a table top. Place one hand at either end of the towel. Push your hands together slowly and watch miniature fold mountains form.

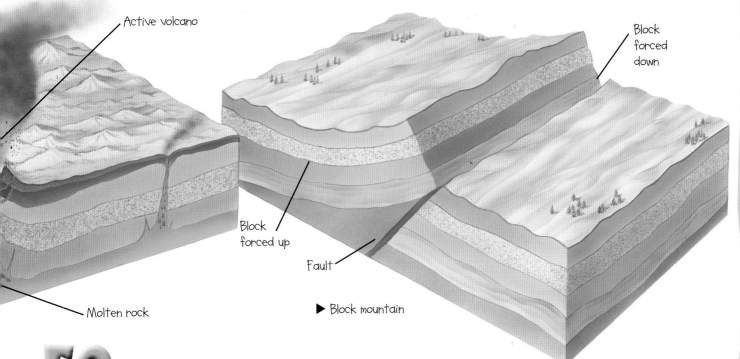

Active volcano

Block forced down

Block forced up

Block forced up

Fault

Molten rock

▶ Block mountain

50 **The movement of the Earth's crust can make blocks of rock pop up to make mountains.** When the plates in the crust push together, they make heat. This softens the rock, letting it fold. Farther away from this heat, cooler rock snaps when it is pushed. The snapped rock makes huge cracks called faults in the crust. When a block of rock between two faults is pushed by the rest of the crust, it rises to form a block mountain.

▲ It takes millions of years for mountains to form and the process is happening all the time. A group of mountains is called a range. The biggest ranges are the Alps in Europe, the Andes in South America, the Rockies in North America, and the highest of all—the Himalayas in Asia.

Shaking the Earth

51 **An earthquake is caused by violent movements in the Earth's crust.** Most occur when two plates in the crust rub together. An earthquake starts deep underground at its "focus." Shock waves move from the focus in all directions, shaking the rock. Where the shock waves reach the surface is called the epicenter. This is where the greatest shaking occurs.

52 **The power of an earthquake can vary.** Half a million earthquakes happen every year but hardly any can be felt by people. About 25 earthquakes each year are powerful enough to cause disasters. Earthquake strength is measured by the Richter Scale. The higher the number, the more destructive the earthquake.

▼ Earthquakes can make buildings collapse and cause cracks in roads. Fire is also a hazard, as gas mains can break and catch fire.

1. Lights swing at level 3

4. Bridges and buildings collapse at level 7

2. Windows break at level 5

3. Chimneys topple at level 6

▲ The Richter Scale measures the strength of the shock waves and energy produced by an earthquake. The shock waves can have little effect, or be strong enough to topple buildings.

Shock waves from the focus

53 **Earthquakes under the sea are called seaquakes.** These can cause enormous waves called tsunamis. As the tsunami rushes across the ocean, it stays quite low. As it reaches the coast, it rises to form a huge wall of water. The wave rushes onto the land, destroying everything in its path.

▲ A tsunami can be up to 100ft (30m) high. The weight and power in the wave flattens towns and villages in its path.

Fault line where two plates rub together

The epicenter is the point on the surface directly above the focus

▲ Focus

Cavernous caves

54 **Some caves are made from a tube of lava.** As lava moves down the side of a volcano, its surface cools down quickly. The cold lava becomes solid but below, the lava remains warm and keeps on flowing. Under the solid surface a tube may form in which liquid lava flows. When the tube empties, a cave is formed.

▲ A cave made by lava is so large that people can walk through it without having to bend down.

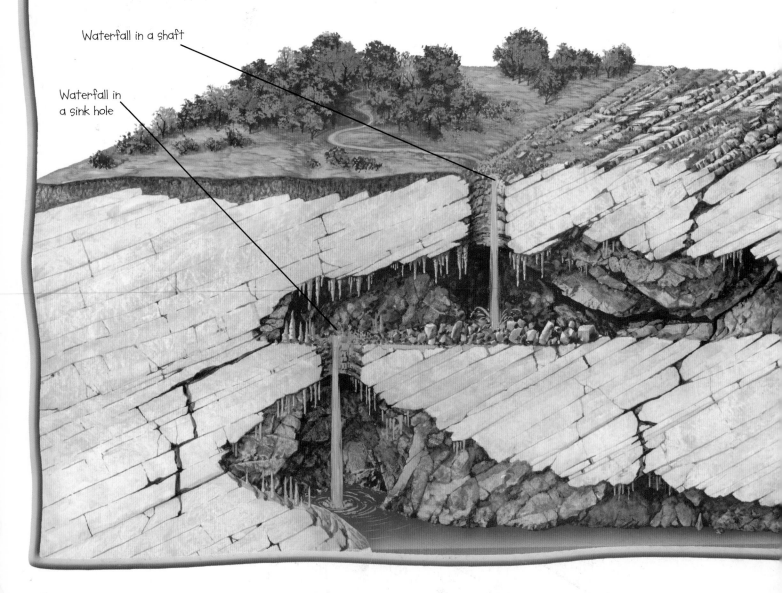

Waterfall in a shaft

Waterfall in a sink hole

1. Water seeps through cracks in rock

55 When rain falls on limestone it becomes a cave-maker. Rainwater can mix with carbon dioxide to form an acid strong enough to attack limestone and make it dissolve. Underground, the action of the rainwater makes caves in which streams and lakes can be found.

▶ Water runs through the caves in limestone rock and makes pools and streams. In wet weather it may flood the caves.

▼ Water flows through the cracks in limestone and makes them wider to form caves. The horizontal caves are called galleries and the vertical caves are called shafts.

2. Underground stream carves into rock

3. Large cave system develops

Gallery

Cave opening

I DON'T BELIEVE IT!
The longest stalactite is 195ft (59m) long. The tallest stalagmite is 105ft (32m) tall.

56 Dripping water in a limestone cave makes rock spikes. When water drips from a cave roof it leaves a small piece of limestone behind. A small spike of rock begins to form. This rock spike, called a stalactite, may grow from the ceiling. Where the drops splash onto the cave floor, tiny pieces of limestone gather. They form a spike which points upwards. This is a stalagmite. Over long periods of time, the two spikes may join together to form a column of rock.

The Earth's treasure

57 Gold may form small grains, large nuggets, or veins in the rocks. When the rocks wear away, the grains may be found in the sand of riverbeds. Silver forms branching wires in rock. It does not shine like jewelry but is covered in a black coating called tarnish.

▲ Gold nuggets like this one can be melted and moulded to form all kinds of jewelry.

58 Most metals are found in rocks called ores. An ore is a mixture of different substances, of which metal is one. Each metal has its own ore. For example, aluminum is found in a yellow ore called bauxite. Heat is used to get metals from their ores. We use metals to make thousands of different things, ranging from watches to jumbo jets.

◀ Silver is used for making jewelry and ornaments.

59 Beautiful crystals can grow in lava bubbles. Lava contains gases that form bubbles. When the lava cools and becomes solid, the bubbles form balloon-shaped spaces in the rock. These are called geodes. Liquids seep into them and form large crystals. The gemstone amethyst forms in this way.

▲ This is bauxite, the ore of aluminum. Heat, chemicals and electricity are used to get the metal out of the rock. Aluminum is used to make all kinds of things, from kitchen foil to airplanes.

▶ Inside a geode there is space for crystals, such as amethyst crystals, to spread out, grow and form perfect shapes.

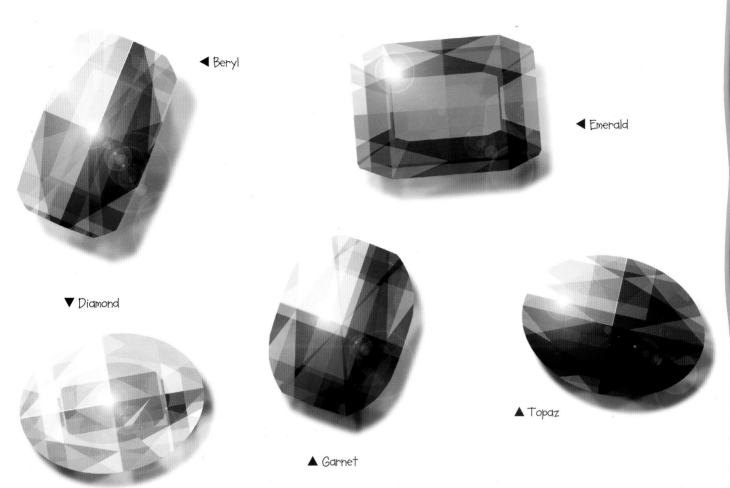

◀ Beryl

◀ Emerald

▼ Diamond

▲ Topaz

▲ Garnet

▲ There are more than 100 different kinds of gemstone. Some are associated with different months of the year and are known as "birthstones." For example, the birthstone for September is sapphire.

60 Gemstones are colored rocks that are cut and polished to make them **sparkle.** People have used them to make jewelry for thousands of years. Gems such as topaz, emerald, and garnet formed in hot rocks, which rose to the Earth's crust and cooled. Most are found as small crystals, but a gem called beryl can have a huge crystal—the largest ever found was 60ft (18m) long! Diamond is a gemstone and is the hardest natural substance found on Earth.

MAKE CRYSTALS FROM SALT WATER

You will need:
table salt
magnifying glass
dark-colored bowl

Dissolve some table salt in some warm water. Pour the salty water into a dark-colored bowl. Put the bowl in a warm place so the water can evaporate. After a few days, you can look at the crystals with a magnifying glass.

Wild weather

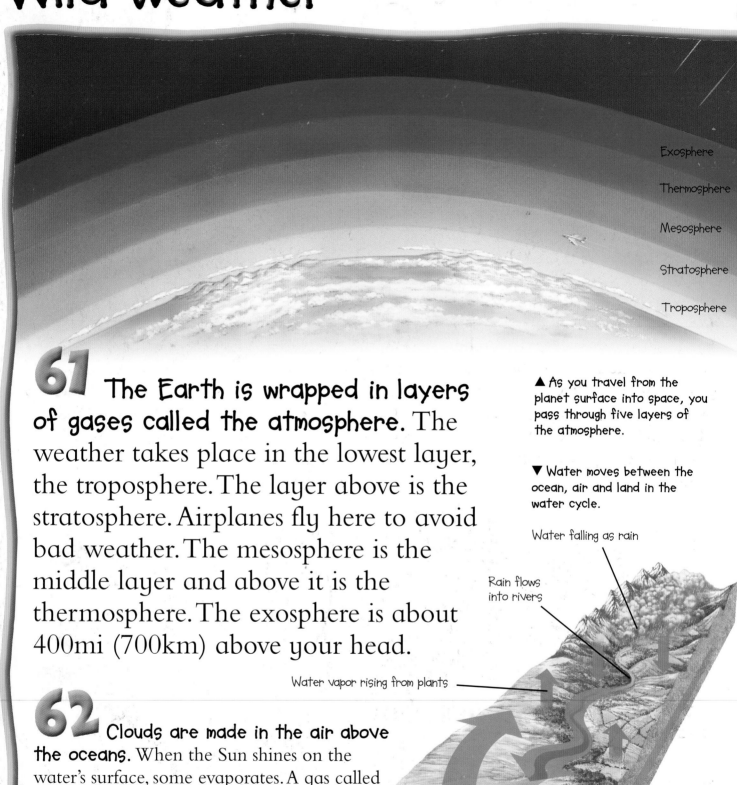

Exosphere

Thermosphere

Mesosphere

Stratosphere

Troposphere

61 **The Earth is wrapped in layers of gases called the atmosphere.** The weather takes place in the lowest layer, the troposphere. The layer above is the stratosphere. Airplanes fly here to avoid bad weather. The mesosphere is the middle layer and above it is the thermosphere. The exosphere is about 400mi (700km) above your head.

▲ As you travel from the planet surface into space, you pass through five layers of the atmosphere.

▼ Water moves between the ocean, air and land in the water cycle.

Water falling as rain

Rain flows into rivers

Water vapor rising from plants

Water vapor rising from the ocean

62 **Clouds are made in the air above the oceans.** When the Sun shines on the water's surface, some evaporates. A gas called water vapor rises into the air. As the vapor cools, it forms clouds. These are blown all over the Earth's surface. The clouds cool as they move inland, and produce rain. Rain falls on the land, then flows away in rivers back to the oceans. We call this process the water cycle.

▶ A hurricane forms over the surface of a warm ocean but it can move to the coast and onto the land.

63 A hurricane is a destructive storm that gathers over a warm part of the ocean. Water evaporating from the ocean forms a vast cloud. As cool air rushes in below the cloud, it turns like a huge spinning wheel. The center of the hurricane (the eye) is completely still. But all around, winds gust at speeds of 200mph (300km/h). If it reaches land the hurricane can blow buildings to pieces.

65 Snowflakes form in the tops of clouds. It is so cold here that water freezes to make ice crystals. As the snowflakes get larger, they fall through the cloud. If the cloud is in warm air, the snowflakes melt and form raindrops. If the cloud is in cold air, the snowflakes reach the ground and begin to settle.

▼ The ice crystals in a snowflake usually form six arms.

64 A tornado is the fastest wind on Earth—it can spin at speeds of 300mph (500km/h). Tornadoes form over ground that has become very warm. Fast-rising air makes a spinning funnel that acts as a vacuum cleaner. It can devastate buildings and lift up cars and traffic, flinging them to the ground.

I DON'T BELIEVE IT!
Every day there are 45,000 thunderstorms on the Earth.

Lands of sand and grass

66 **The driest places on Earth are deserts.** In many deserts there is a short period of rain every year, but some deserts have dry weather for many years. The main deserts of the world are shown on the map.

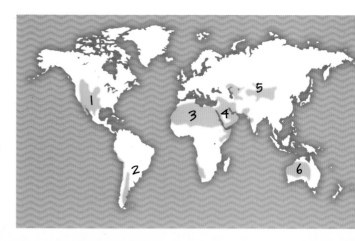

▲ 1. North American deserts—Great Basin and Mojave 2. Atacama 3. Sahara 4. Arabian 5. Gobi 6. Australian deserts—Great Sandy, Gibson, Great Victoria, Simpson.

67 **Deserts are not always hot.** It can be as hot as 120°F (49°C) in the daytime but at night the temperature falls quickly. Deserts near the Equator have hot days all year round but other deserts can have very cold winters.

Ridges of sand being blown into dunes

Barchan dune

Rock beneath the desert

I DON'T BELIEVE IT!

The camel has broad feet that stop it sinking in the sand.

68 **Sand dunes are made by the winds blowing across a desert.** If there is only a small amount of sand on the desert floor, the wind blows crescent-shaped dunes called barchans. If there is plenty of sand, it forms long, straight dunes called transverse dunes. If the wind blows in two directions, it makes long wavy dunes called seif dunes.

69

An oasis is a pool of water in the desert. It forms from rainwater that has seeped into the sand then collected in rock. The water then moves through the rock to where the sand is very thin and forms a pool. Trees and plants grow around the pool and animals visit the pool to drink.

▼ Plants and animals can thrive at an oasis in the middle of a desert.

Oasis

70

A desert cactus stores water in its stem. The grooves on the stem let it swell with water to keep the plant alive in dry weather. The spines stop animals biting into the cactus for a drink.

71

Grasslands are found where there is too much rain for a desert and not enough rain for a forest. Tropical grasslands near the Equator are hot all year round. Grasslands farther from the Equator have warm summers and cool winters.

72

Large numbers of animals live on grasslands. In Africa zebras feed on the top of grass stalks, gnu feed on the middle leaves, and gazelles feed on the new shoots. This allows all the animals to feed together. Other animals such as lions feed on plant eaters.

▼ Three types of animals can live together by eating plants of different heights. Zebras (1) eat the tall grass. Gnu (2) eat the middle shoots and gazelle (3) browse on the lowest shoots.

Fantastic forests

73 There are three main kinds of forest. They are coniferous, temperate, and tropical forests. The main forest regions are shown on the world map opposite.

▲ This map shows the major areas of forest in the world:
1. Coniferous forest 2. Temperate forest
3. Tropical forest

74 Coniferous trees form huge forests around the northern part of the planet. They have long, green, needle-like leaves covered in wax. These trees stay in leaf throughout the year. In winter, the wax helps snow slide off the leaves so that sunlight can reach them to keep them alive. Coniferous trees produce seeds in cones. These are eaten by squirrels.

76 Large numbers of huge trees grow close together in a rain forest. They have broad, evergreen leaves and branches that almost touch. These form a leafy roof over the forest called a canopy. It rains almost every day in a rain forest and the vegetation is so thick, it can take a raindrop ten minutes to fall to the ground. Three quarters of all known species of animals and plants live in rain forests. They include huge hairy spiders, brightly colored frogs, and spotted jungle cats.

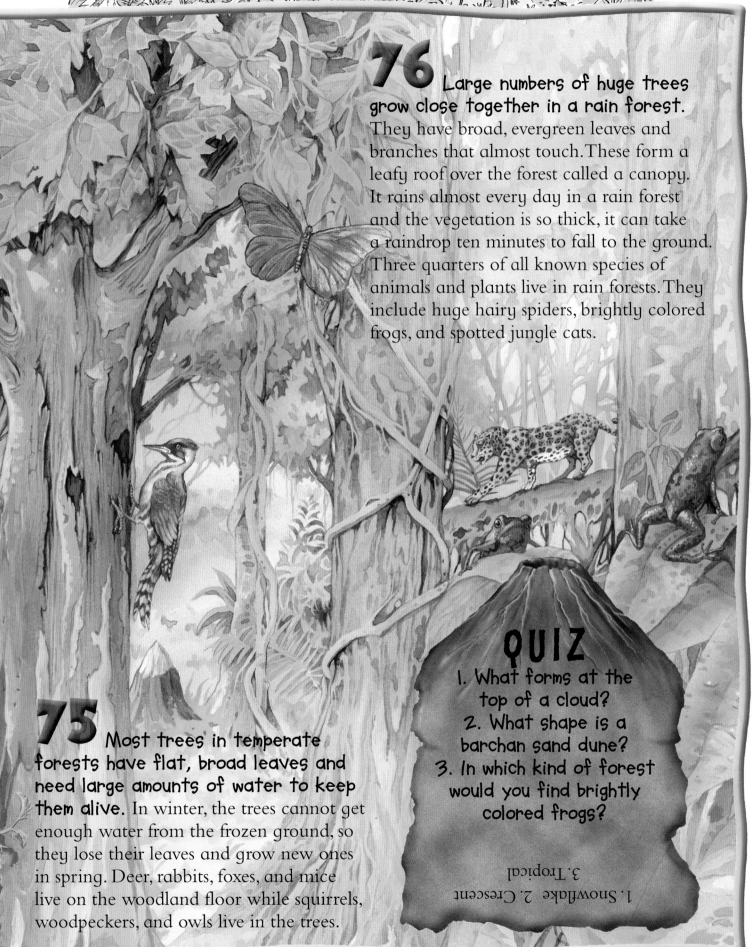

75 Most trees in temperate forests have flat, broad leaves and need large amounts of water to keep them alive. In winter, the trees cannot get enough water from the frozen ground, so they lose their leaves and grow new ones in spring. Deer, rabbits, foxes, and mice live on the woodland floor while squirrels, woodpeckers, and owls live in the trees.

QUIZ
1. What forms at the top of a cloud?
2. What shape is a barchan sand dune?
3. In which kind of forest would you find brightly colored frogs?

1. Snowflake 2. Crescent 3. Tropical

Rivers and lakes

77 **A mighty river can start from a spring.** This is a place where water flows from the ground. Rain soaks into the ground, through the soil and rock, until it gushes out on the side of a hill. The trickle of water from a spring is called a stream. Many streams join together to make a river.

78 **Water wears rocks down to make a waterfall.** When a river flows off a layer of hard rock onto softer rock, it wears the softer rock away. The rocks and pebbles in the water grind the soft rock away to make a cliff face. At the bottom of the waterfall they make a deep pool called a plunge pool.

▼ Waterfalls may only be a few inches high, or come crashing over a cliff with a massive drop. Angel Falls in Venezuela form the highest falls in the world. One of the drops is an amazing 2,664ft (807m).

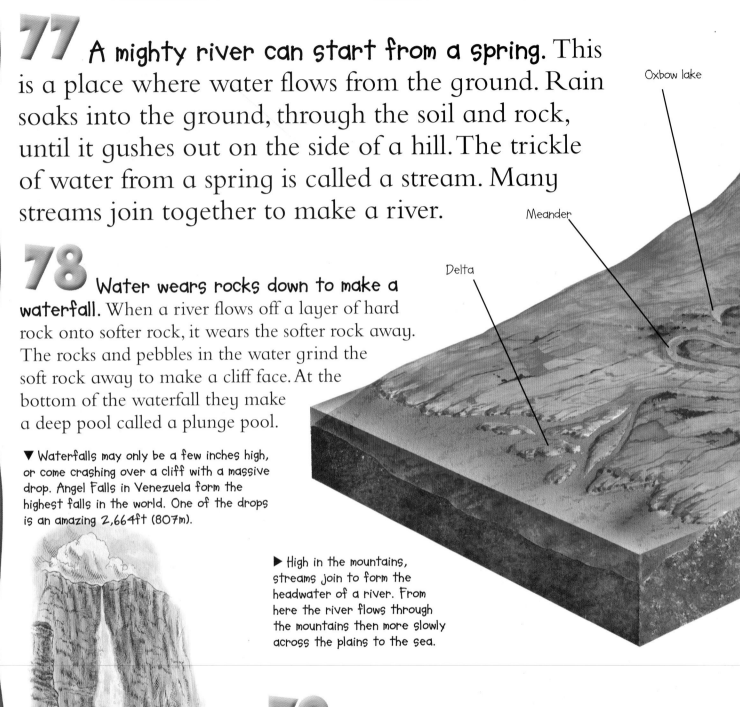

Oxbow lake

Meander

Delta

▶ High in the mountains, streams join to form the headwater of a river. From here the river flows through the mountains then more slowly across the plains to the sea.

79 **A river changes as it flows to the sea.** Rivers begin in hills and mountains. They are narrow and flow quickly there. When the river flows through flatter land it becomes wider and slow-moving. It makes loops called meanders which may separate and form oxbow lakes. Where the river meets the sea is the river mouth. It may be a wide channel called an estuary or a group of sandy islands called a delta.

Headwater

80

Lakes form in hollows in the ground. The hollows may be left when glaciers melt or when plates in the crust split open. Some lakes form when a landslide makes a dam across a river.

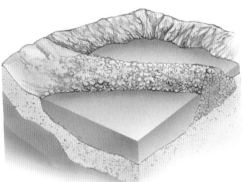

▲ A landslide has fallen into the river and blocked the flow of water to make a lake.

▼ A volcano can sometimes form in a lake inside a crater.

▼ Most lakes are just blue but some are green, pink, red or even white. The Laguna Colorado in Chile is red due to tiny organisms (creatures) that live in the water.

81

A lake can form in the crater of a volcano. A few crater lakes have formed in craters left by meteorites that hit Earth long ago.

82

Some lake water may be brightly colored. The colors are made by tiny organisms called algae or by minerals dissolved in the water.

Water world

83 There is so much water on our planet that it could be called "Ocean" instead of Earth. Only about one third of the planet is covered by land. The rest is covered by four huge areas of water called oceans. A sea is a smaller area of water in an ocean. For example, the North Sea is part of the Atlantic Ocean and the Malayan Sea is part of the Pacific Ocean.

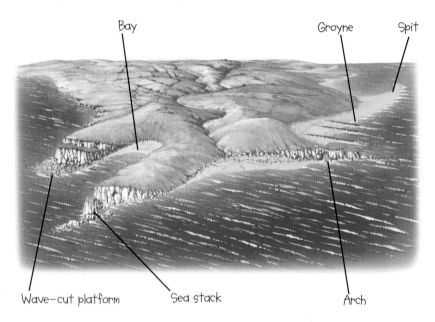

Bay

Groyne   Spit

Wave-cut platform   Sea stack   Arch

84 Coasts are always changing. Where the sea and land meet is called the coast. In many places waves crash onto the land and break it up. Caves and arches are punched into cliffs. In time, the arches break and leave columns of rock called sea stacks.

◄ The rocks at the coast are broken up by the action of the waves.

85 The oceans are so deep that mountains are hidden beneath them. If you stand near the shore, sea water is quite shallow. Out in the ocean it can be up to 5mi (8km) deep. The ocean floor is a flat plain with mountain ranges rising across it. They mark where two places in the crust meet. Nearer the coast may be deep trenches where the edges of two plates have moved apart. Extinct volcanoes form mountains called sea mounts.

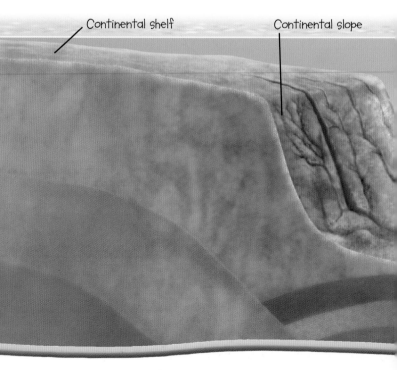

Continental shelf   Continental slope

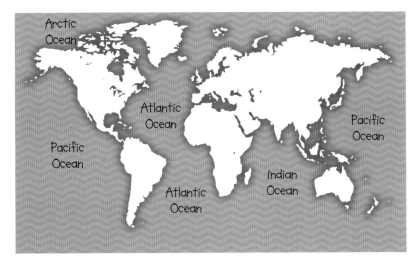

▲ This map shows the major oceans of the world.

87
There are thousands of icebergs floating in the oceans. They are made from glaciers and ice sheets which have formed at the North and South Poles. Only about a tenth of an iceberg can be seen above water. The rest lies below and can sink ships that sail too close.

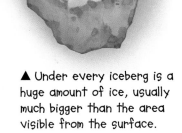

86
Tiny creatures can make islands in the oceans. Coral have jelly-like bodies but they live together in their millions. They make rocky homes from minerals in sea water which protects them from feeding fish. Coral builds up to create islands around extinct volcanoes in the Pacific and Indian Oceans.

▲ Corals only grow in tropical or subtropical waters. They tend to grow in shallow water where there is lots of sunlight.

▲ Under every iceberg is a huge amount of ice, usually much bigger than the area visible from the surface.

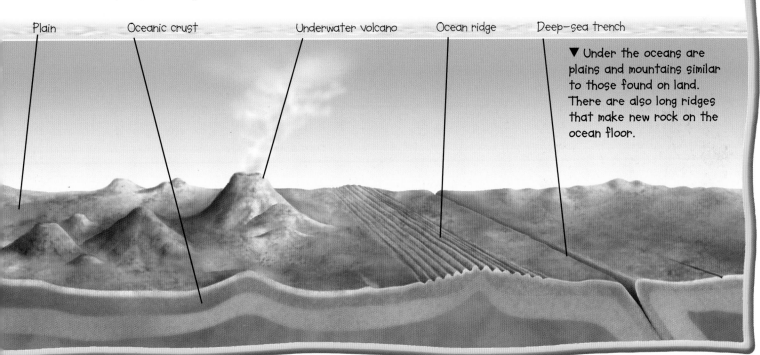

Plain   Oceanic crust   Underwater volcano   Ocean ridge   Deep-sea trench

▼ Under the oceans are plains and mountains similar to those found on land. There are also long ridges that make new rock on the ocean floor.

The planet of life

88 **There are millions of different kinds of life forms on Earth.** So far, life has not been found anywhere else. Living things survive here because it is warm, there is water, and the air contains oxygen. If we discover other planets with these conditions, there may be life on them too.

89 **Many living things on the Earth are tiny.** They are so small that we cannot see them. A whale shark is the largest fish on the planet, yet it feeds on tiny shrimplike creatures. These in turn feed on even smaller plant-like organisms called plankton, which make food from sunlight and sea water. Microscopic bacteria are found in the soil and even on your skin.

▲ Despite being the biggest fish in the oceans, the mighty whale shark feeds on tiny shrimplike creatures and plankton (right).

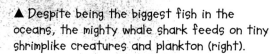

90 **Animals cannot live without plants.** A plant makes food from sunlight, water, air, and minerals in the soil. Animals cannot make their own food so many of them eat plants. Others survive by eating the plant eaters. If plants died out, all the animals would die too.

◄ This caterpillar eats as much plant life as possible before beginning its change to a butterfly.

91

The air can be full of animals. On a warm day, midges and gnats form clouds close to the ground. In spring and autumn flocks of birds fly to different parts of the world to nest. On summer evenings bats hunt for midges flying in the air.

92

The surface of the ground is home to many small animals. Mice scurry through the grass. Larger animals such as deer hide in bushes. The elephant is the largest land animal. It does not need to hide because few animals would attack it.

93

If you dig into the ground you can find animals living there. The earthworm is a common creature found in the soil. It feeds on rotting plants that it pulls into the soil. Earthworms are eaten by moles that dig their way underground.

I DON'T BELIEVE IT!
The star-nosed mole has feelers on the end of its nose. It uses them to find food.

Caring for the planet

94 **Many useful materials come from the Earth.** These make clothes, buildings, furniture, and containers such as cans. Some materials, like those used to make buildings, last a long time. Others, such as those used to make cans, may be used up on the day they are bought.

95 **We may run out of some materials in the future.** Metals are found in rocks called ores. When all the ore has been used up we will not be able to make new metal. Wood is a material that we may not run out of as new trees are always being planted. We must still be careful not to use too much wood, because new trees may not grow fast enough for our needs.

Exhaust fumes from traffic clog up the atmosphere

1. Old bottles are collected from bottle banks

2. The glass or plastic are recycled to make raw materials

3. The raw materials are reused to make new bottles

▲ The waste collected at a recycling center is changed back into useful materials to make many of the things we frequently use.

96 **We can make materials last longer by recycling them.** Metal, glass, and plastic are thrown away after they have been used, buried in dumps and never used again. Today more people recycle materials. This means sending them back to factories to be used again.

Factories pump out chemicals that can cause acid rain. They also dump polluted water in rivers and seas

▼ Here are some of the ways in which we are harming our planet today. We must think of better ways to treat the Earth in the future.

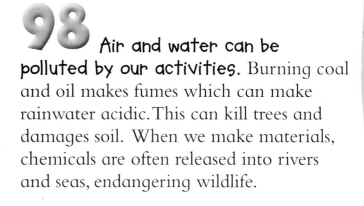

Cutting down trees can devastate forests and wildlife

Trash is dumped in rivers

98 **Air and water can be polluted by our activities.** Burning coal and oil makes fumes which can make rainwater acidic. This can kill trees and damages soil. When we make materials, chemicals are often released into rivers and seas, endangering wildlife.

99 **Living things can be protected.** Large areas of land have been made into national parks where wildlife is protected. People can come to study both plants and animals.

100 **The Earth is nearly five billion years old.** From a ball of molten rock it has changed into a living, breathing planet. We must try to keep it that way. Switching off lights to save energy and picking up litter are small things we can all do.

97 **We use huge amounts of fuel to make energy.** The main fuels are coal and oil, which are used in power stations to make electricity. Oil is also used in gasoline for cars. In time, these fuels will run out. Scientists are trying to develop ways of using other energy sources such as the wind and wave power. Huge windmills are already used to make electricity.

I DON'T BELIEVE IT!
30 to 50 percent of all living species may be extinct by the middle of the 21st century.

Index

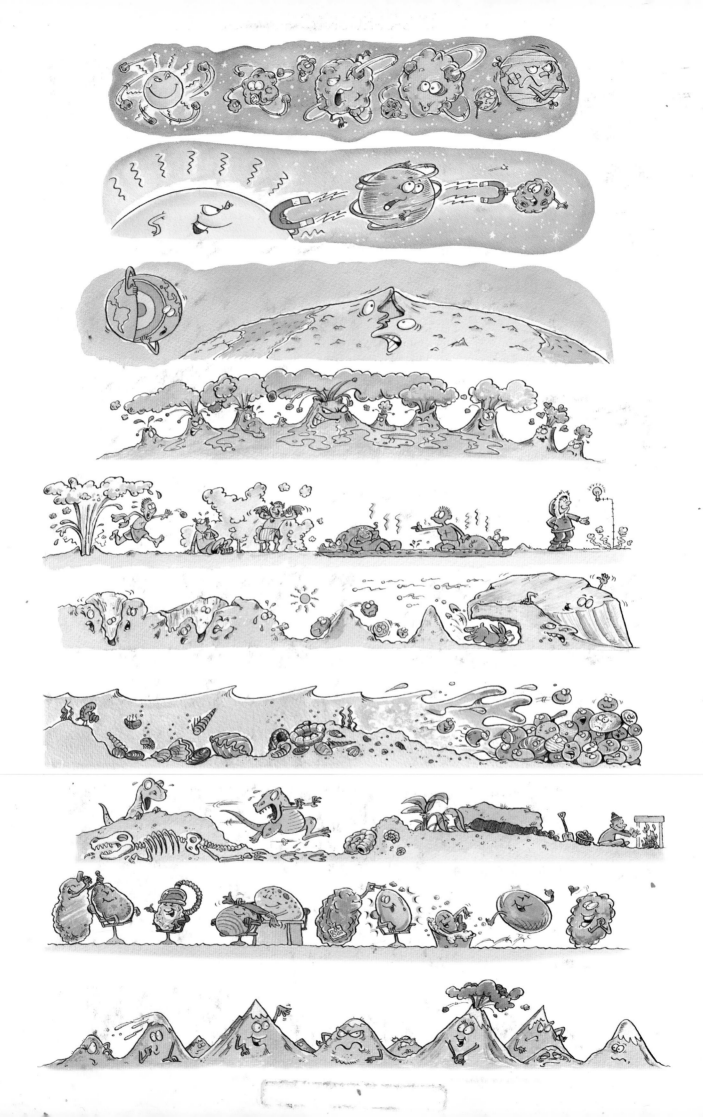